DRAGONFLIES

PowerKiDS press.

New York

Suzanne Slade

With love to my sister, Georganne

Published in 2008 by The Rosen Publishing Group, Inc.
29 East 21st Street, New York, NY 10010

First Edition

Editor: Joanne Randolph
Book Design: Julio Gil
Photo Researcher: Nicole Pristash

Photo Credits: Cover, pp. 1, 5, 7, 9, 11, 13, 15, 17, 19, 21 © Shutterstock.com; pp. 11 (inset), 21 (inset) © Dennis Kunkel Microscopy, Inc.

Library of Congress Cataloging-in-Publication Data

Slade, Suzanne.
  Dragonflies / Suzanne Slade. — 1st ed.
      p. cm. — (Under the microscope: backyard bugs)
  Includes index.
  ISBN-13: 978-1-4042-3819-0 (lib. bdg.)
  ISBN-10: 1-4042-3819-0 (lib. bdg.)
  1. Dragonflies—Juvenile literature. I. Title.
  QL520.S63 2008
  595.7'33—dc22

                                    2006036968

Manufactured in the United States of America

Contents

In Your Backyard

Do you like stories about dragons? Dragons are make-believe animals. There is one kind of real-life dragon that may live in your backyard, though. It is an **insect** called a dragonfly.

Your backyard is a **habitat** full of interesting animals and plants. On a sunny day, you might see a dragonfly zip through the air. If it lands, you can study its colorful body and beautiful wings. Some people are scared of dragonflies because they have a long, pointed tail. Do not worry. This tiny dragon cannot hurt people, and it will not breathe fire, either!

Dragonflies are colorful insects with a long body and big eyes. Dragonflies cannot hurt you, so enjoy this wonderful neighbor in your yard!

Discover the Dragonfly

Dragonflies play an important part in your backyard habitat. Dragonflies eat flying bugs, like **mosquitoes**. They catch lots of insects because they are fast fliers. Dragonflies are not the only hungry animals in your backyard, though. Dragonflies make a tasty meal for birds and frogs. Dragonflies are one part in the **food chain** that keeps your backyard healthy!

People often mistake a bug called the damselfly for the dragonfly. The damselfly has a very thin body, while a dragonfly's body is thick. When a damselfly rests, it pulls its four wings close to its body. The dragonfly rests with its four wings straight out, like an airplane.

Dragonflies beat their wings about 25 times each second to help them go fast.

Many Different Dragonflies

In the United States, there are more than 450 species, or kinds, of dragonflies. There are more than 5,000 dragonfly species in the world. One common species is called the green darner. Its body is 2½ inches (6 cm) long, and it is colored green or blue. Its clear wings measure 6 inches (15 cm) across. A ten-spot dragonfly has dark spots on its wings.

Dragonflies generally live near water. They are found almost everywhere on Earth, except in very cold places. Dragonflies are cold blooded. This means that their body does not make heat. They stay warm by living where it is warm.

This dragonfly is called the widow skimmer. This dragonfly lives in wetlands, ponds, and lakes throughout North America and Mexico.

A Dragonfly Up Close

Like all insects, a dragonfly's body has three main parts. These are the head, **thorax**, and **abdomen**. Two large eyes take up much of the dragonfly's head. In fact, a dragonfly's eyes are so large that they touch each other. Two small **antennae** and a mouth are also on the head.

Just below the head is the short, thick part of the body called the thorax. The dragonfly's four wings and six legs are found on the thorax. The thorax also has tiny breathing holes on it. The thin part of the body that looks like a tail is the abdomen. This is where a dragonfly breaks down its food.

You can clearly see the two huge eyes on this dragonfly's head in this close-up photo. *Inset:* Each huge eye is actually made up of thousands of lenses. This photo was taken by a special tool called an electron microscope.

Magnification: x 43

11

The Life of the Dragonfly

A dragonfly begins its life as an egg and ends it when it dies as an adult. Dragonflies lay their eggs in freshwater, like ponds and lakes. When the eggs **hatch**, tiny **nymphs** are born. The wingless nymphs live and eat underwater. When a nymph grows and becomes too large for its skin, it crawls out of the old skin. This is called molting.

A nymph lives from one to six years in water and may molt up to 15 times. When a nymph leaves the water and molts for the last time, it turns into a dragonfly.

Dragonfly nymphs are also called mudeyes. This dragonfly nymph sits on a leaf in the water.

Making Dragonfly Eggs

When a male, or boy, dragonfly **mates** with a female, or girl, they make a dragonfly egg. The male picks a female based on the color, size, and shape of her wings and body.

Once a male finds the right female, he tries to get her attention. Some males do this by showing off their red backs or white legs. If the male is successful, the female will allow him to hold her head and they will mate. The pair may stay together for seconds or for hours. Then the female will leave to lay eggs.

A male and a female may fly together while mating. When they are done mating, the female will find a wet place to lay her eggs.

Dinner to Go

Dragonflies like to eat small flying insects, such as flies, mosquitoes, and gnats. Large dragonflies may eat butterflies, bees, and even other dragonflies. Dragonflies use their large eyes to find **prey**. Each eye has about 28,000 small **lenses**. Their powerful eyes help them spot food that is up to 40 feet (12 m) away.

A dragonfly often catches prey with its six legs while flying. The dragonfly eats its meal as it continues to fly. Sometimes a dragonfly catches small bugs with its mouth as it flies.

Dragonflies eat lots of bugs that can hurt people and make them sick. This dragonfly is eating a fly.

Enemies of the Dragonfly

Dragonflies are not an easy snack to catch. Their sharp eyes and fast wings help them escape before **predators** can catch them. Birds, spiders, frogs, and other dragonfly enemies must know the right time to catch this tasty meal.

Some dragonflies are eaten when they leave the water for the first time as nymphs to become adults. A frog or bird may pick off a young dragonfly before it knows how to use its wings. A female dragonfly is also in danger when she lays her eggs in the water.

This sparrow has caught a dragonfly. Sparrows are part of the backyard habitat, too.

Wonderful Wings

One of the first things most people notice on the brightly colored dragonfly are its beautiful wings. These four oval wings allow the dragonfly to do many special things.

Before a dragonfly takes its first morning flight, it must sit in the sun and dry the dew from its wings. If a dragonfly feels a need for speed, it can use its powerful wings to fly up to 35 miles per hour (56 km/h). A dragonfly can also hover, or stay in one place in the air. A dragonfly can even fly backward if it wants to!

This photo shows a dragonfly's wings up close. *Inset:* This is a magnified, or much bigger, look at a dragonfly's wings.

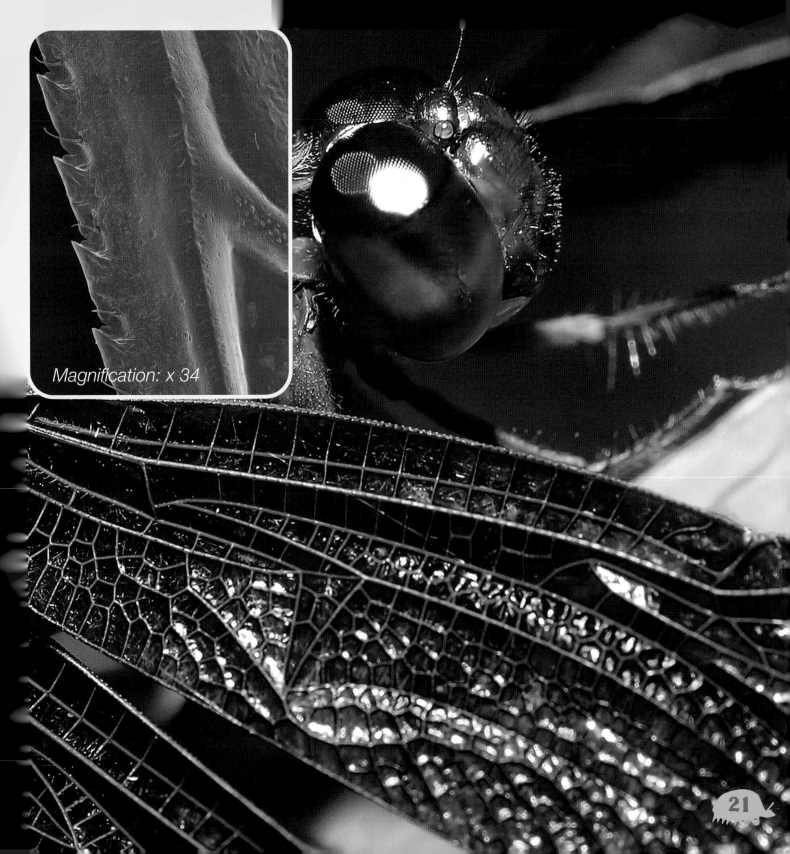

Magnification: x 34

21

Dinosaur Days

An adult dragonfly lives only about 7 to 10 weeks. During this short time, it eats thousands of bugs. It also mates with another dragonfly and leaves behind hundreds of brand-new dragonflies to take its place.

Dragonflies have lived on Earth for a very long time. They were here 180 **million** years ago, when dinosaurs were alive. The dragonflies that lived long ago were larger than the ones we see today. Perhaps dinosaurs watched these colorful dragonflies eat pests like flies and mosquitoes, just as we do today. Even in the days of the dinosaurs, dragonflies were helpful insects!

Glossary

abdomen (AB-duh-mun) The large, rear part of an insect's body.

antennae (an-TEH-nee) Thin, rodlike feelers on the head of certain animals.

food chain (FOOD CHAYN) A group of living things that are each other's food.

habitat (HA-beh-tat) The kind of land where an animal or a plant naturally lives.

hatch (HACH) To come out of an egg.

insect (IN-sekt) A small animal that often has six legs and wings.

lenses (LENZ-ez) The clear, curved parts of each eye.

mates (MAYTS) Joins together to make babies.

million (MIL-yun) A thousand thousands.

mosquitoes (muh-SKEE-tohz) Flying insects that feed on the blood of animals.

nymphs (NIMFS) Young insects that have not yet grown into adults.

predators (PREH-duh-terz) Animals that kill other animals for food.

prey (PRAY) An animal that is hunted by another animal for food.

thorax (THOR-aks) The middle part of the body of an insect. The wings and legs attach to the thorax.

Index

A
abdomen, 10
antennae, 10

B
bird(s), 6, 18

D
damselfly, 6
dew, 20
dragon(s), 4

F
food chain, 6
frog(s), 6, 18

G
gnats, 16
green darner, 8

H
habitat, 4, 6

L
lakes, 12
lenses, 16

M
mosquitoes, 6, 16, 22

N
nymph(s), 12, 18

P
predators, 18
prey, 16

S
species, 8
spiders, 18

T
thorax, 10

Web Sites

Due to the changing nature of Internet links, PowerKids Press has developed an online list of Web sites related to the subject of this book. This site is updated regularly. Please use this link to access the list:
www.powerkidslinks.com/umbb/dfly/